CW00733777

The Open Gate Library 16

TOYS DOWN THE AGES

CHATTO, BOYD & OLIVER

TOYS
DOWN THE AGES

JOHN HORNBY

Illustrated by
VIRGINIA SMITH

For
Frank and Sheila Nettleton

Published by
Chatto & Windus Ltd
40–42 William IV Street
London W.C.2

Clarke, Irwin & Co Ltd
Toronto

© *John Hornby 1972*
First edition 1972

ISBN 0 7011 0469 4

Printed in Great Britain by
Martin's of Berwick

Toys are playthings for children and are meant to give fun.

Although many of them have been educational, especially in recent centuries, the first importance of toys is for pleasure and amusement.

It is not known whether there were toys among the primitive peoples living in caves before history began. Perhaps life was too harsh and demanding to leave time for grown-ups to make playthings or for children to enjoy them. But we all love to play and perhaps these primitive children used natural objects as toys.

Round fruits and stones were probably thrown like balls. Branches of trees could have been the early bats. Many games could have been played with small pebbles. Empty gourds, beaten with sticks, may have made drums. Gourds with seeds in were perhaps the first rattles.

Even tiny baby animals may have been used like our soft cuddly toys.

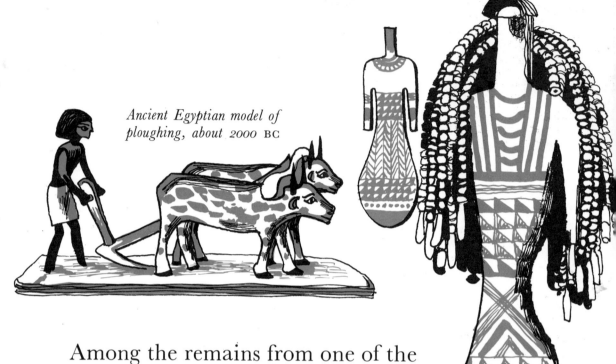

*Ancient Egyptian model of
ploughing, about 2000 BC*

*Two Egyptian paddle
dolls, about 2000 BC*

Among the remains from one of the
early civilisations, Ancient Egypt,
thousands of years BC (Before Christ),
there are many toylike objects.

There are dolls, some of them
"paddle" dolls (so-called because they
are shaped like paddles), models of
boats with people in them, small
figures of servants, workmen, farmers,
sometimes at their work.

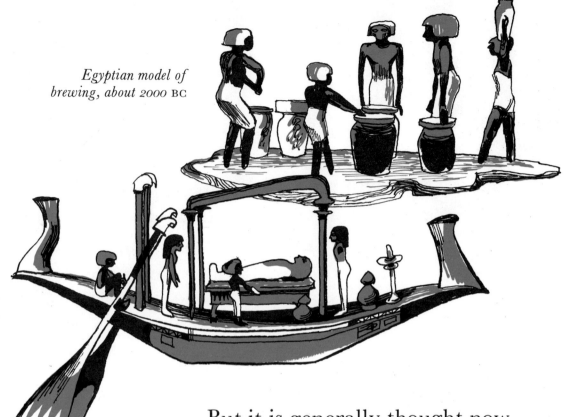

Egyptian model of brewing, about 2000 BC

Egyptian model of funeral boat, about 2000 BC

But it is generally thought now that these were not toys. They were probably models with a magic or religious purpose, buried with people who had died so that they would have help in their next life.

Some of the figures that have been found were probably not dolls but idols to be worshipped.

Egyptian balls of reeds and material, and, at the top, a draughtsman

Later discoveries from Egypt show that they had toys for children too.

Among them were balls of pottery, papyrus, leather, and material bound with reeds; ninepins with marble pins, rattles containing seeds; pointed objects like tops; wooden animals, sometimes with parts that moved; a simple horse on wheels that was pulled along; and some board games.

An early Egyptian board game

Pull-along horse on wheels from Ancient Egypt

Egyptian fish and animal

Toys are not made to last for ever and they are often treated roughly. It is not strange that there were long periods which left no trace of children's playthings.

Our next important groups of discoveries are of the times of the Greeks, a thousand years after Ancient Egypt, and then the Romans.

From Greek terracotta figures of people riding

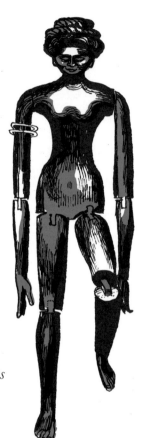

Early Roman terracotta doll,
Roman doll with gold bracelets
and, on right, early jointed
clay doll of Greece

Both Greek and Roman children
had dolls to play with. These were
not the first toy dolls. Much earlier
ones of baked clay have been found
in places like north-west India.

Boy trundling a hoop. From a Greek vase, about 500 BC

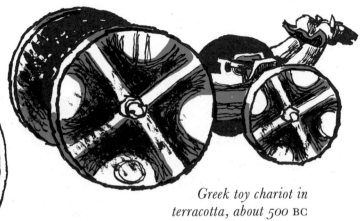

Greek toy chariot in terracotta, about 500 BC

Lady playing the ancient game of knucklebones

We know the Greeks had toys that were made to move by falling sand or mercury, and little chariots and carts that could be pulled along.

They played knucklebones, a hand game, with little bones of animals.

They had bats and balls, whistles, yoyos, hobby-horses, tops and hoops, though the hoops were probably used only by men for exercise.

Roman boys often kept their soldiers in models of the Trojan horse

The Romans had many kinds of toys. The boys had soldiers which were kept in models of the Trojan horse, and they had animals and horsemen of pottery. They also had flat figures, such as Julius Caesar on a horse, which gave the idea for the flat tin soldiers of later ages.

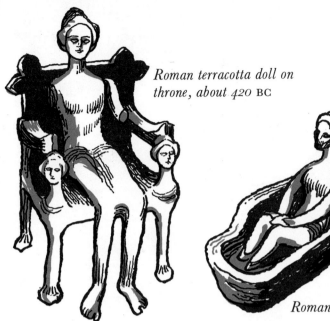

Roman terracotta doll on throne, about 420 BC

Roman doll in bath, about 400 BC

Walking on stilts was popular until recently

Remains of Roman dolls' furniture have been found, and puppets, rattles, dice, balls and marbles. Even the Emperor Augustus played marbles with his slaves. It is still a world-wide game, played also by Zulus.

Stilts have been used by children and grown-ups from Roman times. An old race of Indians danced on stilts to honour their bird god Yaccocahmut.

From a painting of children at play by the Flemish artist Peter Breughel, 16th century AD

The childhood rocking-horse of King Charles I

Rocking-horse of early 19th century

From a Japanese colour print of a child with a hobby-horse

17th-century wooden rocking-horse

A child riding a hobby-horse, from a 16th-century French woodcut

After Roman times came the Dark Ages. Barbarians swept over Europe. We know little about them. Even the Middle Ages which followed left us few records of their toys. Old types went on, but there was little progress.

Children had few years for toys. Most of them had to start work soon after being seven, even in the 15th century. Before that they began when they were younger.

A tricycle

Victorian horse on
wheels, late 19th century

A scooter

Toys were bought from wandering
pedlars or at the seasonal fairs.

One popular medieval toy was the
hobby-horse, which had been known
to the Greeks and Persians. Later it
was replaced by the rocking-horse,
which developed into the horse on
wheels. This in turn gave way to the
tricycle, and to child-size cars.

A modern
child-size
racing car

Trundling the hoop

Whips with two lashes and one lash

A few of the old kinds of toys have disappeared. Hoops and whip-tops lasted into the early years of the 20th century, but are hardly ever seen now—probably through lack of suitable space. But we still have humming tops, which hum or play a tune as they spin.

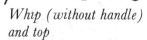

Whip (without handle) and top

From left to right, 19th-century humming top, and grip, peg top with cup, whip top, and modern coloured humming top

English ninepins,
late 19th century

Boy with yoyo,
from a Greek vase

Most kinds of early toys continued throughout the ages, though we may make from plastic today what ancient peoples made from mud, clay and wood.

Yoyos and diabolos came to Europe from China. Ninepins, or skittles, are enjoyed as much by children today as by the Greeks. They are played by grown-ups too, and are called "indoor bowls" in America.

Diabolo,
played with a
light reel on
a cord tied to
two handles

Ninepins or skittles,
from a 17th-century
French illustration

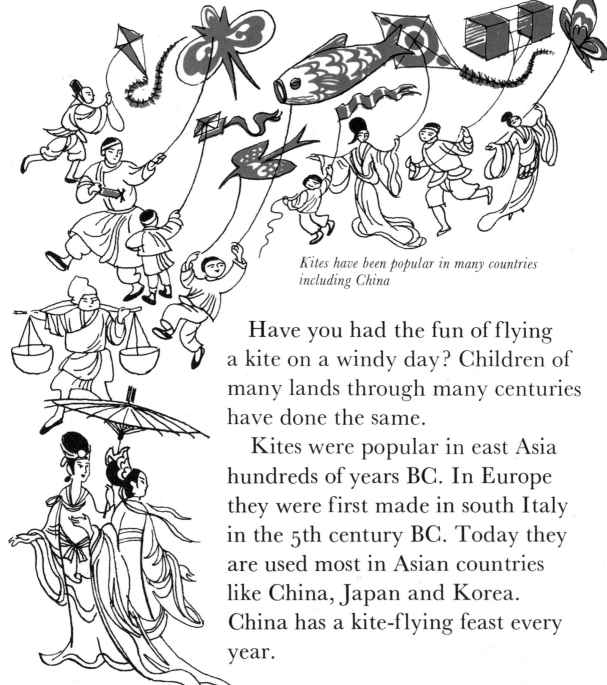

Kites have been popular in many countries including China

Have you had the fun of flying a kite on a windy day? Children of many lands through many centuries have done the same.

Kites were popular in east Asia hundreds of years BC. In Europe they were first made in south Italy in the 5th century BC. Today they are used most in Asian countries like China, Japan and Korea. China has a kite-flying feast every year.

Toy soldiers for boys and dolls for girls—both small imitations of grown-up people—have been beloved toys since earliest times.

Wooden fort made in Britain, early 20th century

For many years the best toy soldiers were made in Germany and France. One child Dauphin (prince) of France had 300 silver soldiers. They have also been made of lead, tin, pottery, wood and cardboard.

18th-century mounted figure

Peasant carving in Czechoslovakia, early 20th century

19th-century French troops, made in modern Italy

Drummer made in 18th-century Germany

*Flat lead Roman model,
about AD 280*

*Solid lead soldier and cannon,
19th century*

Flat, solid, and, later, hollow figures were made, and all kinds of things needed with them to make battle and camping scenes.

Britain entered the industry, and then other countries like America. Figures of all times were sold, as different as medieval knights and Cowboys and Indians. The arms became movable, and soon some parts, like helmets and weapons, could be taken off.

*Models from Mexico, and a
North American cowboy*

*Medieval knights, made in
20th-century England*

Traditional Japanese doll

Early 19th-century German doll

English wooden dolls of the 18th century

17th-century wooden doll

19th-century French doll

18th-century English doll

Dolls are the pride of little girls, who love to play mothers. Yet the early dolls, though called "babies", were small models of grown-ups. Cuddly baby dolls came much later.

Doll-makers were in Nuremberg, Germany, as early as 1413, and in later centuries dolls became more and more varied and attractive.

Modern dolls are very life-like. Many can walk, talk, and go to sleep.

Japan has a Doll Festival each March.

Early 20th-century German doll, fitted to a handle which had a whistle

Fine dolls have come from many countries, such as Germany, France, Britain, America and Japan. Dutch dolls (probably not made in Holland) were wooden. There were dolls of wax, china, rags, celluloid, soft rubber and other materials. Today many are plastic. Some countries like America have made dolls of corn.

Last century saw the start of several dolls' hospitals where dolls could be mended.

Italian doll of papier mâché

Roman rag doll, one of the earliest dolls existing

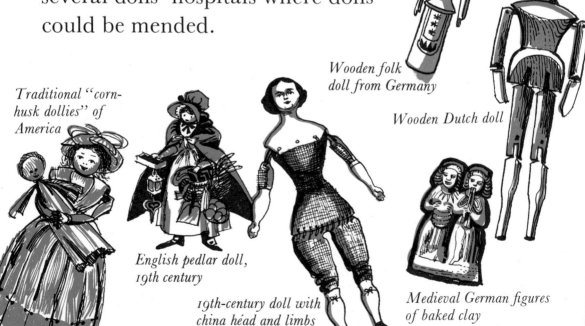

Wooden folk doll from Germany

Wooden Dutch doll

Traditional "corn-husk dollies" of America

English pedlar doll, 19th century

19th-century doll with china héad and limbs and stuffed cloth body

Medieval German figures of baked clay

Russian nesting dolls

Russia is noted for its nesting dolls, which fit inside each other. A Russian visitor gave a set of thirteen to Queen Elizabeth II.

Another kind of doll is the puppet, so-called from the Italian for "doll". Puppets represent humans or animals, and can be made to move and mime a play to spoken words.

Chinese rod puppets from a toy theatre

Marionettes or string puppets

Glove puppet – clown

Glove puppet – jester

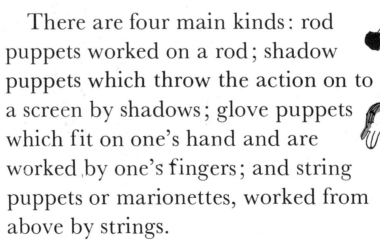

There are four main kinds: rod puppets worked on a rod; shadow puppets which throw the action on to a screen by shadows; glove puppets which fit on one's hand and are worked by one's fingers; and string puppets or marionettes, worked from above by strings.

Before we pass on from dolls we must mention another toy for cuddling, the Teddy Bear, first made about 1904. It was suggested to toy-makers by a photograph of the American president Theodore (Teddy) Roosevelt with a bear cub in the Rocky Mountains, and was named after him.

Teddy Bears, and many other soft animal toys, have been favourites in many countries ever since.

These four animals and the tiger above are modern soft cuddly toys

Early 20th-century American Teddy Bear, one of the first ever made

Baby houses, later called dolls' houses, were first made in Germany and Holland, and to begin with were show cases where grown-ups put expensive miniatures on display.

Then they were given different rooms like real houses, and they were sold as toys for children.

Late 19th-century English dolls' house

18th-century English dolls' house

Some were box-like and simple, but many represented elaborate mansions, complete with small furniture and tiny objects.

In time all kinds of things were made for use with dolls—different clothes, dolls' beds, dolls' prams, delightful tea and dinner sets, little plates of painted plaster food and so on.

*Bed and chairs —
19th-century dolls'
house furniture*

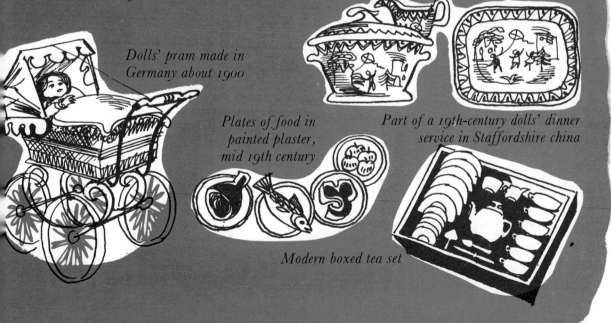

*Dolls' pram made in
Germany about 1900*

*Plates of food in
painted plaster,
mid 19th century*

*Part of a 19th-century dolls' dinner
service in Staffordshire china*

Modern boxed tea set

18th-century Nuremberg kitchen, south Germany

Toy butchers' shops were popular in the 18th and 19th centuries

From Germany also, from the 17th century onwards, came single rooms called Nuremberg kitchens, full of tiny domestic articles.

The same idea was used in small shops, which were popular in the 18th and 19th centuries. Favourites were butchers' and grocers' shops for boys, milliners' and dressmakers' shops for girls.

Noah's Arks and their pairs of animals were always made of wood

Oxcart and owner carved by an American farmer, 19th century

As there had been toy animals from early times, it was only a short step from houses, single rooms and shops to the making of toy farms and farm animals. At first they were of wood, but later sometimes of metal.

But Noah's Ark (with its pairs of animals) was always made of wood, like Noah's Ark of the Bible. When children were not allowed to play normally on Sunday, this was considered a suitable Sunday toy.

Part of a wooden German farm, 19th century

Meanwhile other kinds of toys, like movable ones, had been made. One method of movement was the use of strings tied to parts of the toy to make it move. The earliest we know of this kind is the Egyptian animal with movable jaws.

Sometimes the strings were tied to a hanging weight, which was swung or moved to work the model. Pecking birds were made like this for centuries.

French jumping jack in painted wood, 18th century

Egyptian tiger with movable jaws, about 1000 BC

English pecking bird, 19th century

Child and windmill from a wood engraving, early 17th century

Boy with windmill; from a 16th-century French woodcut

Tumbling dolls worked by moving weights inside the rods, 19th century

19th-century dancer, moved by a jet of steam from the cylinder

Some playthings were moved by the wind. Windmills, usually just plain sails on a long stick, were popular from the 16th century.

Other toys were worked by hidden trickles of sand, or by mercury, water, steam or heat.

Some toys went into action when their balance was changed by a moving weight.

*English clockwork roundabout,
early 20th century*

*Model locomotive,
19th century*

*American wooden doll on
clockwork tricycle, 1870*

The great age of clockwork toys (operated by springs wound with a key) was from the beginning of the 18th century. A movable toy that works by hidden means is fascinating, and seems to have something of magic.

France was very good at making clockwork toys, but by the 19th century England, Germany, America and Japan were making them too. They were sold all over the world.

*English clockwork race on
penny-farthing bicycles,
late 19th century*

Clockwork beetle, Germany, late 19th century

Some of them included musical boxes. Many clockwork toys were singing birds, musicians playing instruments, and dancers accompanied by music.

There were mechanical workers of all kinds, such as knife-grinders, there were acrobats, moving vehicles, clowns, and even performing conjurors.

19th-century conjuror with magic wand

French musical box, with clockwork clowns playing, 18th century

Clockwork boat and sailor, English, 19th century

Clockwork railway engine, tender, and carriage made in Germany, 19th century

Magnetised cat chasing toy mouse

*Magnetised people moving
towards each other*

Another way of making metal toys work was the magnet. Magnetised models of people or animals or insects would move towards other magnetised things—a cat would catch a mouse, a donkey would turn towards a carrot.

Or a magnetised object on top of a box or ledge would move about in answer to the movement of a magnet underneath the surface.

*The magnetised wheels move as the
magnet is moved below the ledge*

19th-century cut-out doll and clothes

*Dissected puzzle,
19th century*

Some movable toys worked by string were sold in cardboard sheets, to be cut out and put together. There were other kinds of cut-outs that made stand-up scenes, or cardboard figures with sets of clothes.

Pictures cut into pieces, often maps of the world, had to be fitted together, and from this idea came our modern jigsaw puzzles.

*Pictures with changeable
heads and faces, early
19th century*

Cut-out figures to make jumping jacks with movable limbs

The most exciting cut-outs were toy theatres, which had their widest popularity in the mid-19th century.

The little stages were of cardboard or wood, and words for the plays were sold with sheets of items to be cut out to make scenery and characters. The sheets were plain black and white or coloured, and the prices were "a penny plain, twopence coloured".

Characters for toy theatres were sold as cut-out sheets

Toy theatres were popular in the 19th century

Many of the plays were from full-size stage ones. Others were from nursery tales like *Jack the Giant Killer* or *The Sleeping Beauty*, or from books like *Oliver Twist*.

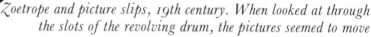

Zoetrope and picture slips, 19th century. When looked at through the slots of the revolving drum, the pictures seemed to move

French magic lantern, 19th century

There were other ways too of looking at scenes and pictures.

Different slides could be slotted into the peepshow for viewing. Flicker books, whose pages of pictures in slightly different positions were flicked over quickly, could give the idea of movement. Later came many inventions like the magic lantern, from which modern cinema films were gradually developed.

The game of cup-and-ball, from a 17th-century print

The many ball games for children of today were not invented until after the discovery in the 19th century that indiarubber could make balls which bounced. Before that, the only bouncing toys came from animal bladders.

But cup-and-ball was popular from early times to the end of last century. The ball was thrown up from a cup, in which it had to be caught again. Even grown-ups, including kings, took their cup-and-ball when walking.

Cup-and-ball sets, early 19th century

A 19th-century Snakes and Ladders board

Early table games were often for grown-ups, but from the 16th century were made for children too. Many are played with counters and dice. Some which were enjoyed in earlier times are still popular today, such as Ludo and Snakes-and-Ladders. There was a children's game called Lotto, which was an early form of Bingo.

An English table game, 19th century

Tiddleywinks, the game of jumping counters, is known as "The Flea" in France.

Our well-loved card games Snap and Happy Families were favourites in the 19th century too. The happy families were named after the trades, Mr Bun the Baker, Mr Tape the Tailor, and so on, and each family, father, mother, daughter and son, had to be collected together.

Tiddleywinks, still played today

Cards from an early 20th-century set of Happy Families

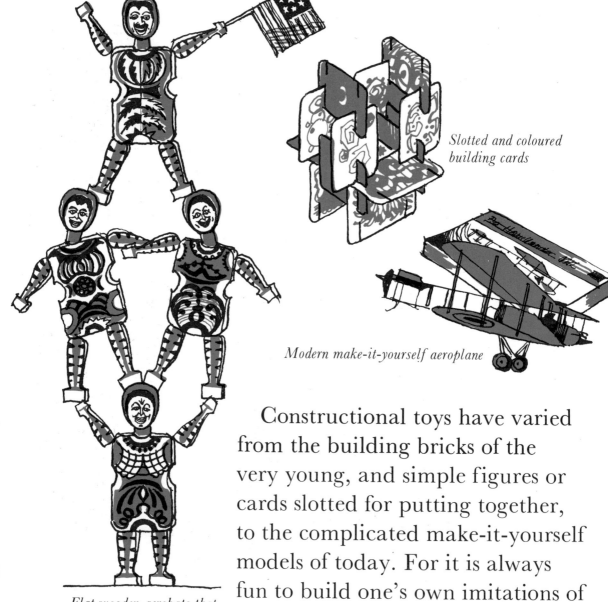

44

Slotted and coloured building cards

Modern make-it-yourself aeroplane

Flat wooden acrobats that could be slotted together in many positions, 19th century

Constructional toys have varied from the building bricks of the very young, and simple figures or cards slotted for putting together, to the complicated make-it-yourself models of today. For it is always fun to build one's own imitations of the surrounding world.

Cargo ship with derrick

Model of the "Mayflower"

Clockwork steamer

Model yacht

American canoe and doll

In fact, the majority of toys are imitations of the real world.
It is natural that many of them are small copies of the great inventions, like ships, trains and motor cars.

Children have sailed things on water all through the ages, from home-made boats of wood or nut shells. Lately we have had all kinds of toy craft, driven by wind, steam, clockwork, rubber bands, chemicals, electricity and wireless control.

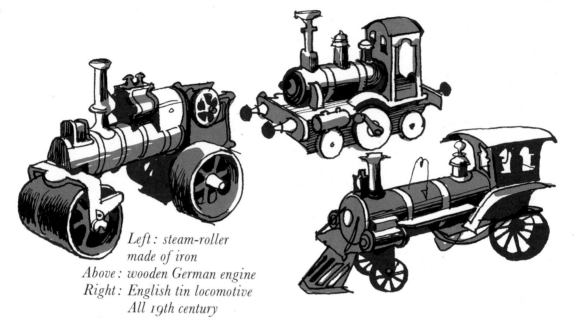

*Left: steam-roller
made of iron
Above: wooden German engine
Right: English tin locomotive
All 19th century*

Modern railway set

The first steam railways were opened in the early 19th century. Small models of the trains were made, and soon toy railways were produced, with all kinds of railway items to go with them, and with the engines worked by clockwork or steam, and later by electricity.

Model railways are a favourite pastime for men as well as boys.

Other vehicles—motor cars, buses, lorries and so on—came soon after the development of the car at the end of the 19th century.

Today tiny matchbox-size models of many makes of cars and lorries are popular. Some lucky children have small cars of their own size in which they can sit and drive.

Some early 20th-century "penny toys"

Wooden pull-along bus, 1925

Clockwork London bus, about 1922

Modern child-size pedal car

*Mechanical aeroplane, early 20th century,
with more modern toys below on left*

The modern aeroplane has been
invented in our own 20th century.
Many different kinds are now copied
for toys. Some of these planes and
gliders are models which can be
flown—worked by rubber bands,
or air currents, or motor.

Now we have toy aeronauts and
moon-landing equipment. Toy
space craft may soon be as popular
as any playthings of the past.

Toy aeronaut